Reaching the American Dream

Kenneth Smith

Author's Tranquility Press
MARIETTA, GEORGIA

Copyright © 2022 by Kenneth Smith.

All rights reserved. No part of this publication may be reproduced, distributed or transmitted in any form or by any means, including photocopying, recording, or other electronic or mechanical methods, without the prior written permission of the publisher, except in the case of brief quotations embodied in critical reviews and certain other noncommercial uses permitted by copyright law. For permission requests, write to the publisher, addressed "Attention: Permissions Coordinator," at the address below.

Kenneth Smith /Author's Tranquility Press
2706 Station Club Drive SW
Marietta, GA/30060
www.authorstranquilitypress.com

This is a work of non-fiction.

Reaching the American Dream/Kenneth Smith – Third Version
Paperback: 978-1-958179-00-0
eBook: 978-1-958179-01-7

*Thank you to my wife, Debbie,
and my son, Jeremy, for your
support and clerical and computer skills.
Thank you.*

Table of Contents

INTRODUCTION 5
YOUR HEALTH... 9
YOUR EDUCATION 13
LANDING A JOB 20
YOUR INCOME 29
YOUR BUDGET 32
YOUR CREDIT.. 41
YOUR HOME 50
SETTING YOUR GOALS...................................... 53
THE FAMILY 57

INTRODUCTION

Reaching the American Dream *The Way I See It* is a series of straight-to-the-point, easy-to-read chapters on how to achieve the American dream.

There is a chapter on health, because without your health, there may not be much you can do. That is why it is important to have health insurance. There is a chapter on education. That is why it is important to stay in school. There is a chapter on how to find a job and a chapter on how much money the average income should be for your work.

There is a chapter on how to make a budget and what should be in it. There is a chapter on loan interest, credit scores, and how much you can afford to borrow. There is a chapter on home buying. There is a chapter on goal setting. There is a closing chapter on the family.

A typical reader should come to understand what the average American dream can look like and use each chapter in this book as a guide in their pursuit to achieving their American dream. Everyone's life is different and unique, this book will help anyone in the search of their American dream.

Before we get started, I think you need to know a little bit about me. I graduated high school in 1970. A lot has changed since then, and people are believing that the American dream will not be achievable for our children and grandchildren. That is why I think that this book may help if we approach being able to achieve the American dream with common sense and a good attitude.

I worked after high school and then joined the navy, knowing that my job was going to be there when I came home in four years, and it was with over four years of seniority.

Therefore, I decided to ask this wonderful lady that I met in Norfolk, VA, while in the navy to marry me. Thank the Lord she said yes, because it will be forty years, we have been together when September 12, 2015, gets here. Believe me when I say that for me, my wife gives me the motivation and drive to do whatever I do. It may not always be easy, but I know that together, we can make it happen. Even when it seems impossible, it becomes possible. I know this to be true, because she was there beside me when I got laid off from my job and then landed my career job as a wastewater plant operator working for the state of Pennsylvania.

Even as a Vietnam veteran, I stayed in the US Navy Reserve and was activated to go to Desert Storm in 1991. I retired from the Navy Reserve in 1995. I think my wife was glad when I did that. I then turned to real estate sales as a second career and did that for around 15 years.

So here I am today. Retired from everything, living in a home that is paid for with the love of my life knowing that our son has a career of his own to build and is working on it. That is the average definition of the American dream. That being said, let's get started and see what it is you have to do to get there.

YOUR HEALTH

The way I see it is that the most important attribute to achieving the American dream is your health. As you read this chapter, you should begin to understand why it is important to do things that can promote good health practices in your lifestyle and why it is important to have good health care to use not only for emergencies but for routine health screenings.

There are organizations out there to help anyone that is not able to work due to injury or illness, such as workmen's comp, social security, and Medicaid, and there are organizations such as United Way where you can get counseling on what to do for your situation. All health issues are protected under the HIPAA Law/Privacy Act.

You know, a person's health is so important that the government will cover a child's health care if they are under the age of nineteen and if you are not under your family's health plan. Insurance companies have to keep you under your parent's family plan until you are twenty-six years old. On the other end of the spectrum, when you turn sixty-five, you become eligible for Medicare. Then, if you still cannot afford health care, there are provisions under the Affordable Care Act that help you get your healthcare coverage.

Insurance companies want you to stay healthy. They may even provide you with a fitness coach to help you lose weight, stop smoking, or even pay for a fitness club membership for you.

Now, if your health is that important to everyone, you have to think twice when you find yourself sitting at a bar after work drinking too much, smoking cigarettes and then driving home too fast because you are late for dinner. And then you might say,

"How is this conducive to staying healthy and helping me achieve the American dream?"–I learned that I was going to have to change after I got married if we were going to get anywhere in our lives and I couldn't be a young carefree sailor anymore. I did stop because I took a look around and realized that most of the successful people I knew did not drink, smoke, or gamble. So by not doing the aforementioned, you will not only have better physical health, you will have better mental health that is less stressful, because your financial health will be more secure.

I can understand that you may want to experiment with tobacco, alcohol, and gambling because they are things you can do legally, and peer pressure will most likely have a role in that, but seriously, that kind of activity is expensive and is very detrimental to your financial health and mental well-being. When I finally stopped smoking and drinking, I seemed to have a little extra money each month, and I think that I felt better after doing so. I also believe that you are most likely not going to meet that special someone in a place where drinking and gambling takes place, if you want to achieve the American dream together and help each other in doing so.

YOUR EDUCATION

Education is most likely the second most important attribute needed to achieve the American dream. Without at least a good high school education, the only way I can think of getting ahead in this world we live in today is most likely by hitting the lottery, being some sort of child prodigy, or having a rich relative leave you money. As you read this chapter, you will come to understand why it is important to stay in school and further your education whenever possible.

Although the aforementioned things are possible, they are not going to happen for the vast majority of us, so let's look at what the vast majority believes will work if we all give it a good, concerted effort.

I will never understand why young people drop out of high school in this age. It does not cost you anything but is worth everything.

The high school diploma is so valuable you can get a job with it that pays more than minimum wage; you can join the military and learn a trade or have a career in the military. You can go to college and be anything you want to be using the GI Bill. The possibilities are almost endless as long as you have that high school diploma. Going to school is so important that if you do not go to school, you are breaking the law and your parents can be in trouble. So keep your parents out of trouble and do the right thing—stay in school.

Remember me saying earlier that a lot has changed since 1970. A high school education was all you needed to get a livable income back then. Today, if that is the extent of your education, it takes two people working to make a comfortable living, and if one of you lose that job, you can lose the income you need to live comfortably.

My dad was born in 1921. He only went to school to the eighth grade. I, at least, accomplished the same thing, but I needed a high school diploma. My son was born in 1977, and he has an associate's degree. And it seems like he is having a hard time getting by at times. I can only imagine what it would be like if my son had dropped out of high school. So please stay in school and try to do whatever it takes to get some kind of training or degree that is going to help you be successful in whatever it is you think you may want to do with your life. No one ever said life is easy, but you can do whatever you think can help you make it easier, if you try. Your high school or college guidance counselor can help you. On your way, don't be afraid to ask for help. That is their job to be there for you.

You can do all kind of things that can help you further your education along. The best one I can think of is joining the military. You can use your GI Bill. You may even achieve some college credits during your enlistment and being a veteran with a college degree makes you more employable. But you are more employable just because you are a veteran, and your employer may send you to school for a trade such as plumbing, electrical, HVAC, or commercial driving. You could have picked up some of this training in the military or not.

Any way you look at it, if you can show a prospective employer that you have some post-secondary education and have experienced some of what it takes for making a more mature and able-bodied person for today's job market, it is a big plus for your ability to succeed in it.

Do not get discouraged if the military won't work for you and make it possible to further your education. Do not give up. There are programs out there in the way of grants and student loans sponsored by our government that can help you achieve your

American dream. Do not give up if you think, this isn't for me. Try and go for it anyway because this is the greatest country in the world, and everyone has an equal opportunity to a good life, and it will start with a good education. That is why we have laws that say this will happen, if you try. Laws such as affirmative action and equal opportunity employment give everyone a chance to become successful. That is what makes our country great. It can happen. Our president can tell you so.

This part is going to probably help you and your parents make your education happen. There are two things I need you to think of. Number one is find out if your parents are going to be able to help you with school. Maybe they have been saving all along for you to go to school. But if it is you with young children, I would go see a financial adviser about opening a 529-college plan for them. Every state is different, but they give you tax advantages for saving for your children education. I also would encourage your children to work and save for their education, if they have a summer or after school job or work while in college. It will give them work experience and help pay for school. Sometimes, even college students need to know that college is not high school. It doesn't just happen.

Choosing a career that you are going to invest in by paying to go to school, requires some thought and research as to how much school and the demand for the type of work within the geographical area that you may choose to live in. In other words, you need to know if it will be possible to find work within the area you choose to live in and if you are willing to relocate to be able to advance your career to gain experience if need be.

Most colleges and universities know whether the availability for employment is strong or not, as they probably would drop a poor course of studies from their curriculum, due to poor employment aspects and low student enrollment. Even if employment aspects are in high demand, they may be low in some locations, so you have to be able to go where the work is.

As an example, you know that you want to be an architect and there are six architectural firms where you want to live. You have that brand new degree, but the demand is full in your hometown. You may have to move out of town to get some experience and build your resume and hopefully return home to one of the hometown firms someday.

LANDING A JOB

Landing that job that will allow you the ability to live and achieve the American dream is what your parents or guardians want for you, and it is what America wants for you also. So, let's see what kind of money we need to make and what we have to do to get the job.

You can choose to go to college, but you don't have to if you don't choose to or feel that you can't afford to, but it is a good thing to consider and to look into just to open that door of opportunities as wide as it can go, if at all possible. Now you have to remember that without that high school diploma, the door of opportunities only opens just enough to see the light.

Your career does not define who you are. You define who you are. Your job defines what you do. If you want the job, you have to show any prospective employer that you are a healthy, educated, clean-cut individual willing to do what it takes to learn the position you are applying for. Now, remember your first impression will be a lasting impression as the interviewer considers you over others for any open position. You are not as good as you can be. You are only as good as others perceive you to be at this point.

Forget about landing a job. You have to land an interview first. That usually starts with writing a resume to post or present to an employer along with his application for employment. That is so you can land an interview. So, tell the truth and list good character references. Past employers are usually good if you had worked in high school or college. They may call theses references before they decide to interview you. You may find help in putting a good resume together at your local employment agencies or if you are in school, there may be a course that you can take in doing this.

When you land that interview, dress for success. Remember, look the part as though you can step right in and do the job. Pay attention to how people are dressed when you turn in your application. That is right. Don't just mail it in or e-mail it. Hand-carry it, if possible, and introduce yourself to the person who takes your information. When your information does not have a stamp or e-mail address in it, it makes you look more of an ambitious type of individual and you may land that interview. When you do land that interview, do not be late; be early.

You can ask questions that you may know the answer to so that your reply can be, "That is what I thought," "That is why I applied for this position. Because I have training or experience performing whatever it is that you asked about." I would not recommend asking the starting wage unless that information is nowhere to be found and the interviewer did not provide it. Then, be discreet when asking that question if it does not come up or you still haven't known. Also, ask if there is health care and if there is an employee contribution required.

There are many occupations that you can do by yourself without being declared as an employee as such. You can be considered as a self-employed contractor or maybe a salesperson. You may be a carpenter, a truck driver for yourself, a realtor, or a car salesperson working for commission. Don't forget that it will be up to you to report your income so you can report your tax liability. Then save money so you can pay your taxes. You have to realize that in this type of work, you should be a good speaker because you will be looking for new work all the time. It may seem like you are preparing to be interviewed all the time instead of filling out a job application. You may be writing ads to get your name out there. You may want to take a course in public speaking and ad writing.

You may be the type of person that may want to start your own company or business. This will take a lot of planning, and you may have to borrow money to get started. You will most likely need some business degree to get that accomplished. That will mean that you will need to work with the small business association to make a business plan so you can seek financing. Then there is the hiring of a work staff and the need to pay them sometimes before you even make any money. But if you go to school and learn how to do it, you can have a successful business.

We have talked about interviewing for a job, working for yourself, and starting a business, but you also have to consider what it is that you are going to be doing or want to do. There has to be a need for your occupation that will last long enough to retire from with a good pension. I don't think

I would go to work for a company that all they did was make incandescent light bulbs when you know they are being faded out of production. Why is that so important? Just go out and find a new job, right. I believe that you need to have that career job at least by the time you are thirty years old so you have enough time to build that retirement by the time you reach sixty-five years old. I always thought that it takes twenty years to grow up, thirty years working and thirty years to enjoy retirement. Anymore then that is a plus, give or take a few years. For the military or college, maybe a period of unemployment.

There are some people that start their careers at a young age in the military or maybe law enforcement. This type of employment awards you the ability to retire early usually after twenty years. This can give you time to start a second career.

When you are out there looking for that career job, you have to decide what you would like to pursue, whether it be public sector or private sector employment, or it could be in an occupation with ties to both, such as the ship building industry or aviation industry, even the automobile industry. Or any industry that may be under contract with any governmental authority whether it is federal, state, or local even your school districts.

That is why the people that we vote for can have such a vital role on our employment and careers. Every US citizen has the right to vote, so please get registered and vote your opinion on election days.

I always believe that if I had a public sector career job or a job that was under contract with any governmental authority that my career would be pretty secure, and I could spend my money in the private sector and that is what makes the economic wheel go around. This scenario must be true. Just go to any military town like Norfolk, VA and imagine what it would look like if the US Navy was not there. I rest my case.

There is only one other thing I can think of that is important to consider when pursuing your career occupation and that is whether you believe that it would behoove you to work in a unionized environment or in a non-unionized environment. The answer to that question use to be a no-brainer. Union employees always seem to enjoy a higher standard of living to the point that nonunionized companies tried to emulate so as to not lose their good employees to a union job. Some people seem to believe that unions are losing their effectiveness. I don't think I would turn a good job down just because I was required to pay union dues if I knew that I didn't have to think about things such as health care, sick leave, paid vacation, and worker safety because all that stuff is worked out at the negotiation table through the collective bargaining agreement, and I can vote on it or I can even get my self-involved and even help negotiate the agreement. Unions are not just something you pay into. They are what the members make them into. That decision is going to have to be left up to you alone.

YOUR INCOME

How do you know if the job you are applying for is going to meet your needs? You can research things like salary for most jobs. At the US Department of Labor Employment estimates, on your computer, today it states that the average wage in today's workforce is $22.00 an hour.

A wage of $22.00 an hour may sound good but do not forget that, that is only average. Some jobs pay less, and some pay more. That is why education and training is so important. But even at $22.00 an hour, there are a lot of taxes and insurances that have to come out of that money like what is listed below.

1. Income tax (15%) = $3.30
2. State income tax (3%) = .66
3. Local income tax (2%) = .44
4. Social Security (6.2%) = $1.37
5. Medicare (1.45%) = .32
6. Union dues, if needed (1.5%) = .33
7. Retirement, if required 6.5%) = $1.43
8. Health care, if required (5%) = $1.10

Total = 8.95

$22.00 - $8.95 = $13.05 net an hour or $2, 262.00 a month

Now, anyone can see that a tight budget may need to be required if there is no other income. I will be using this amount in "Your Budget" chapter.

Inflation is determined by the things we use and buy every day. This rate of inflation runs around 3% a year on average. As of the first publication of this book, back in 2014, the average income across the country was around $22.00 an Hr. As of today, it is safe to say that by adding 3% a Yr. to. $22.00 amortized, we are now looking at $26.27 an Hr. or more if inflation is higher than 3%. Therefore, inflation has to be considered, as you can see how fast the cost of living increases.

YOUR BUDGET

A budget is not something that is hard to do. You just have to do it or you will lose track of what you are spending, and you will not have enough money to pay your bills when they come due. Your budget is something that you need to review just before every payday just to ensure that you are staying on track, because your bills have to be paid on time. Not when you think you can pay them. Let's take a look at how easy it is to do.

The only thing you need to know to make a budget is how much money you have and what it is you need to spend it on. Let's say that you have a net income (after taxes, health care contribution, social security, union dues, Medicare, and unemployment insurance, don't forget your retirement contribution) of $2,262.00 a month, which is around $22.00 an hour for 40 hours a week.

The only thing you need to do is think about what you know that you need to spend such as food, gas, electricity, telephone, car payment, mortgage or rent. Don't forget car inspection or your mother's birthday when they occur. Make a list of all of your known expenditures on one side of a sheet of paper and get a total, then put your known income on the other side of the paper and subtract your expenditures from your income and hope you have some leftover.

You already gave up over $1, 550.00 in payroll deductions a month. That is why you only have a net income of $2, 262.00 a month. Before you can take care of any significant other person in your life, you have to be able to take care of yourself, and there are five things you are going to need. So let's take them out of the equation and look at what they may cost. The first thing is food and water. Food for one person is $200.00–$300.00 a month. Let's split the difference and say $250.00 water and sewage fees although being paid by the quarter will usually be around $50.00 a month, if you divide by three months.

Housing is not something that is not inexpensive. A two-bedroom apartment can cost $500.00–$800.00 in an average small town in the US. In some big cities, they can go from $1,000.00–$2,000.00. That is why being able to make an average wage is so important. I would recommend buying a house when possible. The average home in America according to Zillow is between $150,000–$160,000. If you were to mortgage a home at $150,000 for 30 years at 4% interest, that would be $716.00 a month plus taxes and insurances being put in escrow. You are looking at $1,000.00 a month. For this budget, we will split the difference between $500.00 and a $1,000.00 and make housing for $750.00.

Now that you have a home to live in, you have utilities to pay for. We already talked about water and sewage. We were talking about food and water but there is gas, electric, and garbage. Add those three up. You are looking at $200.00 a month on a monthly average depending on where you live. You most likely want a cell phone, cable TV, internet, and a home phone. Bundle all that into one bill and you can be looking at another $200.00.

Now let's look at your way to and from work. A good used car today will run you $10,000.00. To finance $10,000.00 at 4% for five years and if you add the tax and title work on to the payment, you are looking at $200.00 a month. Then there are tires, inspections, oil changes, and car insurance. I can say that, that is a good $50.00 a month just to keep up with all that stuff plus your driver licenses, and registration renewal, and it is not unusual to spend $200.00 a month on gasoline.

You need to save for health-care co-pays and deductibles. No one wants to get sick or have an accident. You have to take care of your health because without it, there isn't much you can do. I would save at least $25.00 a month for this.

Last, but not without saying, you need to put money away for your clothes. I am going to say $40.00 a month. You will need life insurance just in case the inevitable happens, say $35.00 month.

When we add up our basic needs and subtract from what we have, we will know if your income is going to be strong enough just to take care of yourself. So let's do it.

1. $250.00 Food
2. $50.00 Water and Sewage
3. $750.00 Housing
4. $200.00 Gas, electric, and garbage

5. 5. $200.00 Phone, internet, cellphone, and cable TV
6. $200.00 Car payment
7. $50.00 Maintenance and car insurance
8. $200.00 Gasoline
9. $25.00 Health care co-pays and deductibles
10. $40.00 Clothes
11. $35.00 Life insurance

Total = $2000.00

Wow, you did it and you have $262.00 a month left over for yourself, so don't blow it on cigarettes, booze, or gambling, because smoking just a little over a pack a day, that $262.00 is gone; if you go to your favorite bar every day and spend $10.00, you will be $100.00 short every month, and if you do both, you will be $362.00 short every month. Gambling is a no-brainer. The gambling business is not in the business to give you money.

Try taking that $262.00 and find that significant other by going out to dinner or a movie. And if you both have that average income, you will live and reach the American dream pretty easy. This may not happen right away, so you may have to get a part-time job to help yourself out, especially if you are paying off a student loan or a car loan that may be more than $200.00 a month. You have to look at your budget every day so that you keep it fresh in your mind. You need to keep a checkbook ledger. It is easy to lose track of your money in today's high-speed world using debit and credit cards to pay for everything.

There are other things that go along with living the American dream. They are having the ability to use your paid vacation, buy a new car every ten years or so, and save for a rainy day in case you have to fix that car, because it is getting old. Or, it is a possibility that you get laid off from work for a while or you are trying to save for your children's education or whatever it is you might want to save for, you have to put it into a savings account and make a budget so you know what you have it earmarked for and how much you need to put in to be able to reach your goal. Then you can figure out how much you need to save and treat that just like a bill. Believe me, if you don't do this, trying to live and achieve the American dream can seem like a struggle. This is where that part-time job or that other income source in the family helps.

I would think that when you get your household and savings budgets figured out and you know that you have enough income to support your spending and savings, build your savings up so as to have a surplus of three months' income that you can always have in reserve, just in case you do have a period of unemployment by any of your income sources or there may come a time that you may have to take a break in employment due to an illness or injury.

I hope you can see why this budgeting process is so important, but I want you to know that it is not really as hard as it may seem when you are reading this chapter. But as you do this and review it on a routine basis and keep track of it, it will become a habit, such as brushing your teeth, and you will know your financial situation like the back of your hand at any point in time, and this will help you make the right decisions when you decide to spend your hard-earned income.

<u>YOUR CREDIT</u>

In other chapters, we talked about your health, your education, and your job, how much money you might need, how to budget that money, but this chapter is about your credit. You need to understand credit in order to maintain good credit. You can borrow money for a home or car cheaper with good credit than you can with bad credit.

You need to establish credit before you can build good credit. Most utility companies will extend you some credit because that is how they sell you their product by your use per month. You can also get a credit card and pay it off at the end of each month. This will result in you achieving a good credit report and a high score on that report if you pay your bills on time.

Creditors or lenders report all credit that they have extended to you under your social security number to the credit bureau of which there are three—Trans Union, Equifax, and Experian. This information is compiled and will determine what your FICO credit score will be from 300–850. You should try to keep your score above 700.

Your credit score is not the only thing that will determine whether you will get any credit extended to you. Your work history and income have a lot to do with your credit. In most cases, a lender will want to know if you have been working at the same type of occupation for at least two years. It may not have to be with the same employer, but it is good if it is. Because a long-term employer can tell them if you will most likely remain a long-term employee and that you reported your wages to them accurately. They will figure out your debt-to-income percentage and that cannot be any more then say 38% of your gross income. I would keep it lower than that so you have a little wiggle room, as people might say, because we are only talking long-term debt, such as school loans, car payments, credit cards, child support, furniture payments, and a mortgage, if you have one. Lenders know that the rest of your income goes to living expenses. So, let's take a look at that $22.00/hr job.

$22.00 × 40 (hrs) = $880.00 a week × 52 (wks) = $45,760.00 ÷ 12 (mos) = $3,813.33 gross income a month

1. School loan = $75.00
2. Car loan = $225.00
3. Credit cards = $50.00
4. Mortgage = $1000.00

Total = $1350.00

$1,350 ÷ $3,813 = 35.4%

There you have it. Your debt to income is almost at the maximum. A good rule of thumb is to never pay on your transportation that is more than one year's income and never buy a home that more than three years' income. So, in this scenario, if you are paying for two cars, neither one of them should be worth more than $22,000.00. Your home should be worth only around $135,000.00.

Although this last thing is not part of your long-term debt, it has to be considered as such. It is your savings. Because you need three sources of income to be able to retire. First, whatever your employer provides you with such as a defined pension plan that you may pay into or a 401-K plan. Second, your social security, and third, some sort of savings plan, most likely an IRA of some sort. The first two sources of income will meet your basic needs but the third is what you use to fight off inflation and makes it possible to live and enjoy your retirement with little financial concerns and enjoy the American dream without having to go to work after your retirement.

So there you have it, it is possible to make it on $22.00 an hour in today's America. That is where my goal would be set or even higher to make the American dream my achievable goal.

Whenever you seek credit, I think it is important to know how much it cost to borrow money and how to keep that cost down, so let's take a look at the biggest source of credit most everyone has in their household. That is a mortgage. The only parts of a mortgage that you have to look at for this scenario are the principal and interest. The taxes and insurance you have to pay no matter what. So let's look at the principal and interest of a one-hundred-thousand-dollar mortgage.

Let's say that we borrowed $100,000 at 4% for thirty years. Your payment each month will be $477.42. A month without the taxes and insurance for 360 months is $171,871.20. As you can see, your loan will cost you just under $72,000. Now imagine what you could do with just a third of that $72,000. That's right; that is $24,000. Now think, what it is that you would do with an extra $477 a month? That might be just what you need to help send your children to college or buy a motor home and do a little traveling. All you need to do is pay around $40 a month extra on your mortgage and say use your income tax return each year toward the principal of your loan. You will reduce your loan by years and save a lot of money in interest. Buying a home can give you a higher income tax return, so make that money work even harder for you by putting it down on the principal of your mortgage.

The only other thing I can say is pay attention to the interest rate at which you are borrowing money, because if you borrowed that $100,000 at 5% instead of 4%, your monthly payment will be $536.82. You may not be able to put that extra $40 a month on the principal of your mortgage. Make it one of your long-term goals to maintain a good credit score and use your credit cards wisely. You will be able to borrow money at a lower interest rate.

The interest rate is important, but something called the APR is usually higher than the interest rate. The difference between the APR rate and the loan rate is the cost of the procurement of your loan, so make sure that the difference between the two is not too high when shopping for a loan. This will help you shop for a loan before you allow a lender to run your credit. Your credit score will stay higher. Having your credit run will lower your score a little each time. This is especially important when looking for a mortgage.

There is one other thing that is a loan product out there known as an ARM or adjustable-rate mortgage. I, as a former realtor, cannot really see anyone buying the home that they are going live in with an adjustable-rate mortgage, because that type of mortgage will not give you the ability to know what your payment is going to be. That will make it hard to do a budget. This is what got so many people in trouble during the great recession of 2008–2009. Most of the borrowers were not totally made aware of that happening. They only looked at the monthly payment at the time the mortgage was taken out. Not what it was going to be in the future. So, be careful what you agree to. Try to go with a fixed rate mortgage. If you want to adjust your interest rate, you can refinance your mortgage. Although the reasons for doing this may vary, I don't think I would do this unless my interest rate dropped at less by two percentage points due to the closing costs involved with doing so.

I just mentioned closing costs associated with a mortgage. A very few people understand all the fees, but, by law, you have to have them explained to you. No one can buy something without knowing what it will cost you and that is what you are buying—a mortgage. Your realtor will explain those costs to you, and have you sign a written document saying that you understand them, and your loan officer will do the same. Your closing officer or attorney will ask for copies of these documents. You may ask the seller to pay some of that cost for you and that can happen if the seller wants a fast sale but not all the time. Therefore, be prepared to bring a cashier's check to the closing table to cover down payment and cost. This could be in the thousands. You will know that by that estimated cost document that your realtor and loan officer had you sign, and your closing agent or attorney will let you know how much to bring to the table in the form of a cashier's check.

YOUR HOME

There may be some good reasons not to buy a home. For example, you live and work in New York City and real estate is just too expensive. Another place like this may be Washington, DC. It really doesn't cost a whole lot more to buy a home then rent one. There are government programs to help you do just that. Veterans can buy homes with no money down. Lower-income people can buy with very little down and get the government to help with interest and closing costs. You can ask these questions when you talk with a good realtor. You can even get the government to subsidize the interest on your home, if your income is low under the USDA. If you build in a rural area, see a realtor. This is why buying a home is associated with the American dream.

Homeownership is part of the American dream. It is most likely the average person's main source of wealth, and there are so many things you can do with your home besides living in it. You can deduct the interest that you pay on your mortgage off of your income, so you don't have to pay tax on it. That is like getting Uncle Sam to help pay for your house even if he is already helping you in any of the other ways I have mentioned above.

You can borrow money against your home and that money may be tax deductible also. You can even use it as part of your retirement if you want to take what is known as a reverse mortgage out on it. Anytime you borrow money against your home, it is money you can deduct the interest you pay on it from your income and don't have to pay tax on it. You can use the money for almost anything. So you can see how your home can be your most important asset.

I learned that the average homeowner would buy three homes over twenty-four years. That is an average of one every eight years and that will keep the value of your home up. For instance, say you have a brother, and your father gives each of you the identical house built on your own acre of land; both homes were appraised at $50,000 at the time. You sell your home for $60,000 and buy a home in the suburbs for $75,000 and pay it off in fifteen years. That is pretty easy to do. You only borrowed $15,000. Your brother stayed in his home at 3% inflation a year. Your home is worth $113,444 and your brother's is worth $90,305. This is why homeownership is a good investment that is hard to lose on.

SETTING YOUR GOALS

Goals are things that that people set when they know what they want to accomplish. For example, say you would like to lose fifty pounds, so you know what it is that you want to accomplish. That is to be a long-term goal. So now, set the term, let's say in one year. Take a look at your progress every month and make adjustments, if necessary, to keep yourself on track to achieve your goal because you know that your goal is achievable in a year's time. So, every day, you have to ask yourself what it is you are going to do to help yourself achieve this goal, that is, not doing the things that can hinder you and achieving this goal. So, by setting goals, you have to think about the choices you have to make every day. For this example, a person may choose to count calories or exercises by taking a walk or choose to take the stairs instead of the elevator.

We all have the ability to choose and by setting goals for yourself, you can ask yourself why you want to accomplish this goal to lose fifty pounds. You will see that it will most likely be linked to another goal that you want to accomplish, such as your health, your ability to do a job, or find that special someone in your life that can help you both achieve the American dream together.

So, by setting goals for yourself, you can see just by thinking about them each day you will find yourself doing things that help you facilitate achieving them and not doing things that will hinder you. This will make your decision-making in life right most of the time.

If I had to grow up and build a career that would take me to my American dream, I think I would set my goals high, review them regularly to see if I am on track, and think about them every day so I can do the things that will keep me on track and avoid doing the things that won't and make the decisions necessary to achieve my goals.

Prioritizing your thoughts and goals are important. Also, it can be so easy to get off track if you don't do the things that have the most significance to you before you start letting your wants override your needs. You may have a strong desire to go see this rock concert that is coming into town. You just got paid; you need groceries and gasoline, and you have to put some money away to pay your gas, electric, and phone bill. You should have made being able to go to that concert a short-term goal but not a major priority. This would enable you to maybe cut back on your daily sodas or not go out to eat for a few weeks and save up to go to the concert. There will always be someone or something out there that will try and entice you to stray away from your goals or disfigure your priorities. Please do not take the attitude of

"I am working. I have the right to be able to do whatever it is that I want to do." Because all work and no play is not fun you will have fun if you plan for it. Set a goal, prioritize it, and you will be better off for approaching things in that manner.

This is the best way I know to never let my wants take a higher priority over my needs.

THE FAMILY

I won't even try to give any advice on holding a family together. I am not a marriage counselor or anything like that, but I do believe in what I have written in this book. I also believe that in order for your marriage to succeed, you have to be able to take care of yourself before you can take care of others. I can only hope that this book will help you live, teach, and achieve your American dream. Even though statistics show from the US Census Bureau that a typical marriage only last eight years, and if you look at other statistics out there, that there is less than a 50 percent chance that any marriage will last in the United States today. Therefore, all I can say on this topic is to communicate with one another and use this book as a guide to live and teach your children from and you and your whole family will have a better chance of achieving the American dream.

There is only one other thing that I can think of and that is that one of the best ways to learn is by example. My mother was a good cook, and as a result, we always had one meal together a day in the home I grew up in. That meal was lunch, because my father worked second shift, and even in elementary school, I could walk home for lunch, and we would have that meal as a family. Because my mother was a good cook, we had company quite often. The home I grew up in was always full of family and friends on the holidays. So, think of it this way, a family that eats together stays together. Therefore, get yourself a good cookbook and try to find recipes that your family can gather around the table for. You will save money by not going out to eat, and it will build a stronger family.

www.ingramcontent.com/pod-product-compliance
Lightning Source LLC
LaVergne TN
LVHW020438080526
838202LV00055B/5251